XTREME AIRCRAFT

GLIDERS

A&D Xtreme
BOLD HI-LO NONFICTION
An imprint of Abdo Publishing
abdobooks.com

S.L. HAMILTON

TAKE IT TO THE XTREME!

GET READY FOR AN XTREME ADVENTURE! THE PAGES OF THIS BOOK WILL TAKE YOU INTO THE THRILLING WORLD OF GLIDERS. WHEN YOU HAVE FINISHED READING THIS BOOK, TAKE THE XTREME CHALLENGE ON PAGE 45 ABOUT WHAT YOU'VE LEARNED!

ABDOBOOKS.COM

Published by Abdo Publishing, a division of ABDO, PO Box 398166, Minneapolis, Minnesota 55439.

Printed in the United States of America, North Mankato, MN.

092021

012022

THIS BOOK CONTAINS RECYCLED MATERIALS

Editor: John Hamilton

Copy Editor: Tamara L. Britton

Graphic Design: Sue Hamilton

Cover Design: Laura Graphenteen

Cover Photo: Shutterstock

Interior Photos: Alamy-pgs 32-33 & 38-39; Alexander Schleicher-pg 35; DG Flugzeugbau-pg 44; DreamsTime-pg 27 (inset); FAI-pg 20 (inset); Getty/iStock-pgs 13 (inset), 20-21, 26-27, & 28-29; Lennart Batenburg-pg 36; Magnus Einarsson-pgs 40-41; National Soaring Museum/1-26 Association-pg 37; Nordic Gliding & Aviation/Jens Trabolt-pg 34; Science Source-pgs 12-13; Shutterstock-pgs 1, 4-5, 10-11, 14-15, 16-17, 18-19, 22-23, 24-25, 30-31, & 42-43; Smithsonian Institution-pgs 6-7; US Air Force-pgs 8-9, Wright-Brothers.org-pg 8 (inset); Yorkshire Philosophical Society-pg 6 (inset).

LIBRARY OF CONGRESS CONTROL NUMBER: 2021943544

PUBLISHER'S CATALOGING-IN-PUBLICATION DATA

Names: Hamilton, S.L., author.

Title: Gliders / by S.L. Hamilton

Description: Minneapolis, Minnesota : Abdo Publishing, 2022 | Series: Xtreme aircraft | Includes online resources and index.

Identifiers: ISBN 9781532197352 (lib. bdg.) | ISBN 9781098219550 (ebook)

Subjects: LCSH: Aviation--Juvenile literature. | Gliders (Aeronautics)--Juvenile literature. | Sailplanes (Aeronautics)--Juvenile literature. | Hang gliders--Juvenile literature.

Classification: DDC 629.1333--dc23

TABLE OF CONTENTS

CHAPTER 1

GLIDERS

Gliders are aircraft that fly on rising warm air currents without the use of an engine. There are several types of gliders, including paragliders, hang gliders, and sailplanes. Daring pilots soar in the skies, hearing only the sound of the wind and nature in their ears.

XTREME FACT

NASA's space shuttles, which flew from 1981 to 2011, are the largest and heaviest gliders ever built.

A Discus PH-1162 sailplane soars quietly above a nature area.

CHAPTER 2

GLIDER HISTORY

Scientist and inventor George Cayley (1773-1857) developed the first **fixed-wing**, heavier-than-air glider. In 1853, the Cayley Glider soared with an adult pilot through the skies for 180 meters (591 ft) at Brompton Dale, England.

George Cayley was the "Father of Aviation."

Otto Lilienthal (1848-1896) studied bird flight to further understand how to build a flying machine. From 1891-1896, he made nearly 2,000 short flights in 16 different glider designs. The pioneering German aviator was nicknamed "The Flying Man."

Otto Lilienthal died on August 10, 1896, after crashlanding in his glider. One of his gliders is part of America's National Air and Space Museum collection.

Wilbur Wright flies a Wright Glider in 1901.

Americans Orville and Wilbur Wright created three Wright Gliders from 1900-1902. The brothers made hundreds of glider flights, gaining more knowledge each time. In 1903, they attached a motor to a glider and made history by taking the first controlled flight of a powered airplane on December 17.

As motorized planes took to the skies, gliders continued to be used for fun, in sporting events, and even by the military. During **World War II**, single-use gliders transported troops and heavy equipment. A transport plane towed the sailplane, then released it to quietly glide down to land near the action.

A US Air Force Waco CG-4A is towed through the skies. It was the most-used troop and cargo glider of World War II.

CHAPTER 3

PARAGLIDERS

Paragliders are lightweight aircraft that fly when wind fills the soft, fabric wing. Pilots sit in a harness or speed bag, suspended by lines below the wing.

XTREME FACT

On October 10, 2019, three Brazilians soared 588 kilometers (365 mi), setting the distance paragliding world record. Flying as a team, Marcelo Prieto, Rafael Saladini, and Rafael Barros each spent 11 hours in the air.

Wind makes it easier to launch a paraglider, but even without a breeze, the pilot can launch just by running.

A paraglider is launched and landed using the pilot's legs. To launch, the wing is spread out on the ground. The pilot makes sure the lines are untangled, then locks into the harness. The pilot holds the lines and runs. The wing fills with air and begins to lift.

XTREME FACT

In case of falls or crashes, paraglider pilots prefer to take off from a soft surface such as sand or grass. They also wear helmets for protection.

A paraglider lands by pulling the brake lines. Experienced pilots can land at precise locations.

Most paraglider distance races are 50 or 100 kilometers (31 or 62 mi).

Paragliders fly at speeds of 20-75 kph (12-47 mph). Weather conditions, **thermals**, the type of paraglider, and the pilot's weight determine speed. Paragliders fly in cross-country races and in aerobatic stunt challenges called acro paragliding .

In aerobatic competitions, paraglider pilots do spirals, rolls, wing-overs, stalls, and more. Their tricks are very dangerous and require a high skill level.

CHAPTER 4

HANG GLIDERS

A hang glider is like a big kite capable of carrying a person.

XTREME FACT

Early hang gliders did not have harnesses. Pilots hung onto the bar of the glider with their arms. They were told to only go as high as they dared fall!

Hang gliders are light fabric stretched over an aluminum frame with the pilot in a harness below. The sail, or wing, is about 9 meters (30 ft) wide. To direct the glider, the pilot is suspended in an A-frame and shifts their body left, right, or pulls forward or backward.

Air needs to be moving across the wing's surface at about 24-40 kph (15-25 mph) to generate lift.

Hang gliders are usually foot launched. The pilot runs down a slope until the air lifts them up. Hang gliders may also be tow launched. A vehicle, boat, winch, or a microlight plane gives them a lift. To land, the pilot **stalls** the glider by pushing the control bar as far out as possible. The nose tips up, the glider slows, and the pilot lands upright on their feet.

With the glider's nose up, a pilot starts his landing.

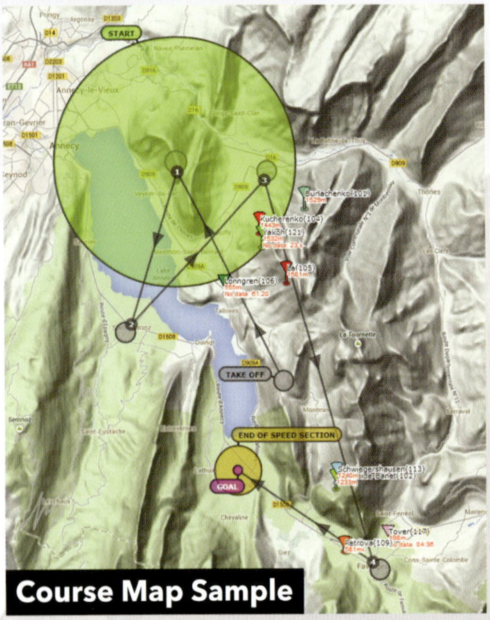

Course Map Sample

Glider cross-country race-to-goal competitions have pilots riding **thermals** while trying to follow a specific course. They have a finish line or goal to reach and must land in a specific area. The fastest time wins.

Skilled hang glider pilots can stay in the air for hours. As of 2020, the airborne record was more than 9½ hours.

CHAPTER 5

SAILPLANES

A sailplane has **fixed wings** and a **fuselage**. There are several classes of sailplanes, with most based on wing size or the weight of the plane. Wingspans usually range from 13.5-30 meters (44-98 ft).

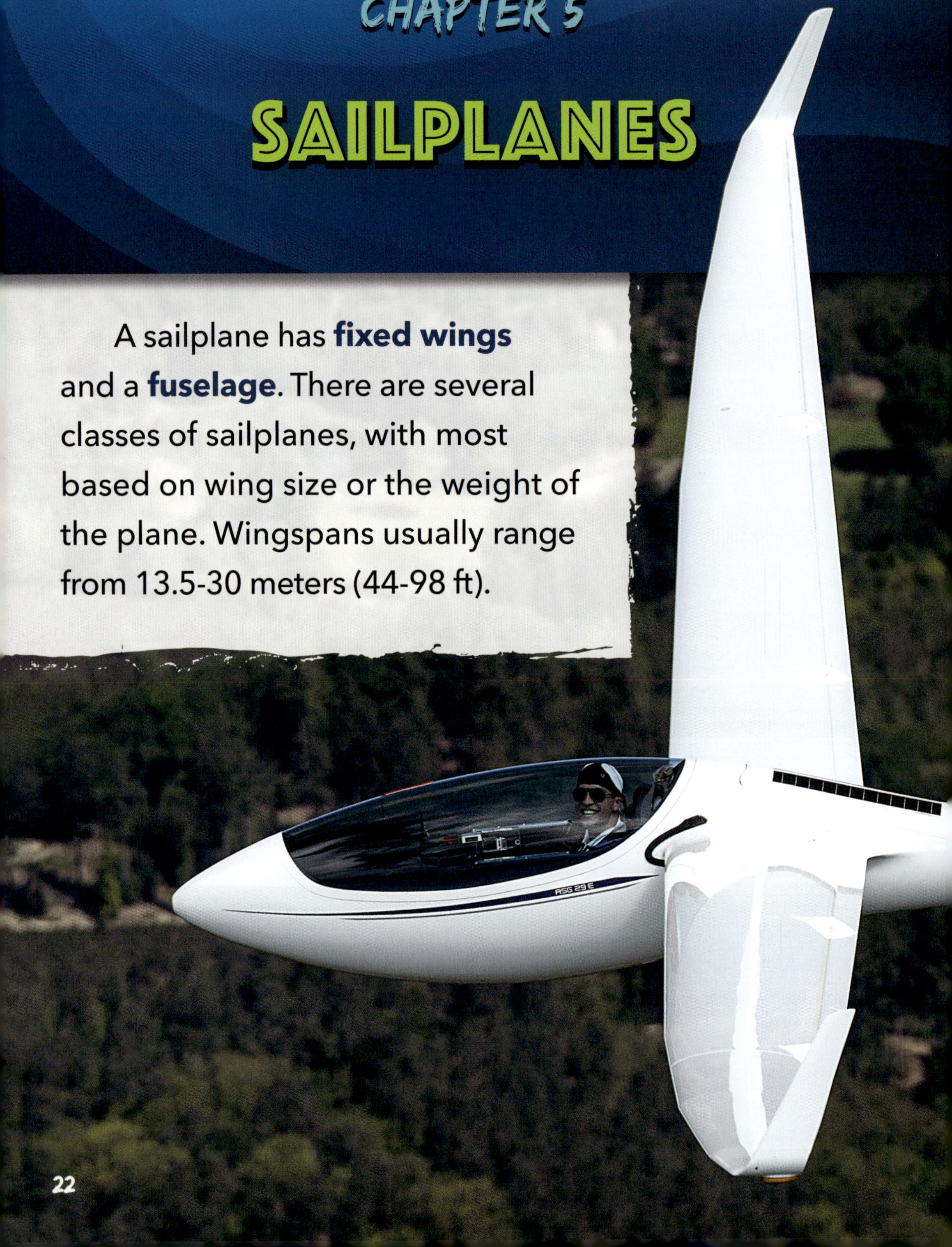

XTREME FACT

On January 15, 2009, Captain Chesley "Sully" Sullenberger credited his vast experience as a glider pilot with being able to safely land a disabled Airbus A320 on the Hudson River. It was called the "Miracle on the Hudson." Sully saved all 150 passengers and 5 crew members.

Today's sailplanes are made of very strong, lightweight materials such as carbon fiber, fiberglass, and Kevlar.

The advantage of an aerotow is that the higher the launch, the longer the sailplane will stay in the air.

Sailplanes may be launched several ways. An aerotow is one of the most common. The glider is towed behind a powered aircraft. At about 3,000 feet (914 m), the cable is released and the sailplane flies on its own.

An auto-tow launch works similar to an aerotow, but with a truck towing the glider. Rope is used with a tow hook. The vehicle speeds down an airstrip, launching the glider into the air.

During a winch launch, the glider is attached to a long cable. The winch's powerful motor quickly rewinds the cable, pulling the glider up into the sky.

Winch launches are made into the wind. A pilot must find a source of lift quickly to gain altitude.

For a bungee launch, the pilot moves the glider backward, stretching an attached length of elastic tubing. When there is enough tension, the plane is released, slingshotting into the air.

A winch launch costs the pilot much less than an aerotow.

Self-launching gliders are equipped with an engine that is powerful enough to allow the plane to take off on its own. Many have the engine and propeller located behind the cockpit. Once the plane is in the air, the launching equipment **retracts** into the **fuselage** and the plane soars without power.

A self-launching glider takes off. Once in the air, the engine and propeller fold back down and seal in their own compartment.

A glider lands much like a regular plane. Smaller landing gear, often a single wheel, is placed directly below where the pilot sits. Landing gear may stay in place or **retract** into the plane.

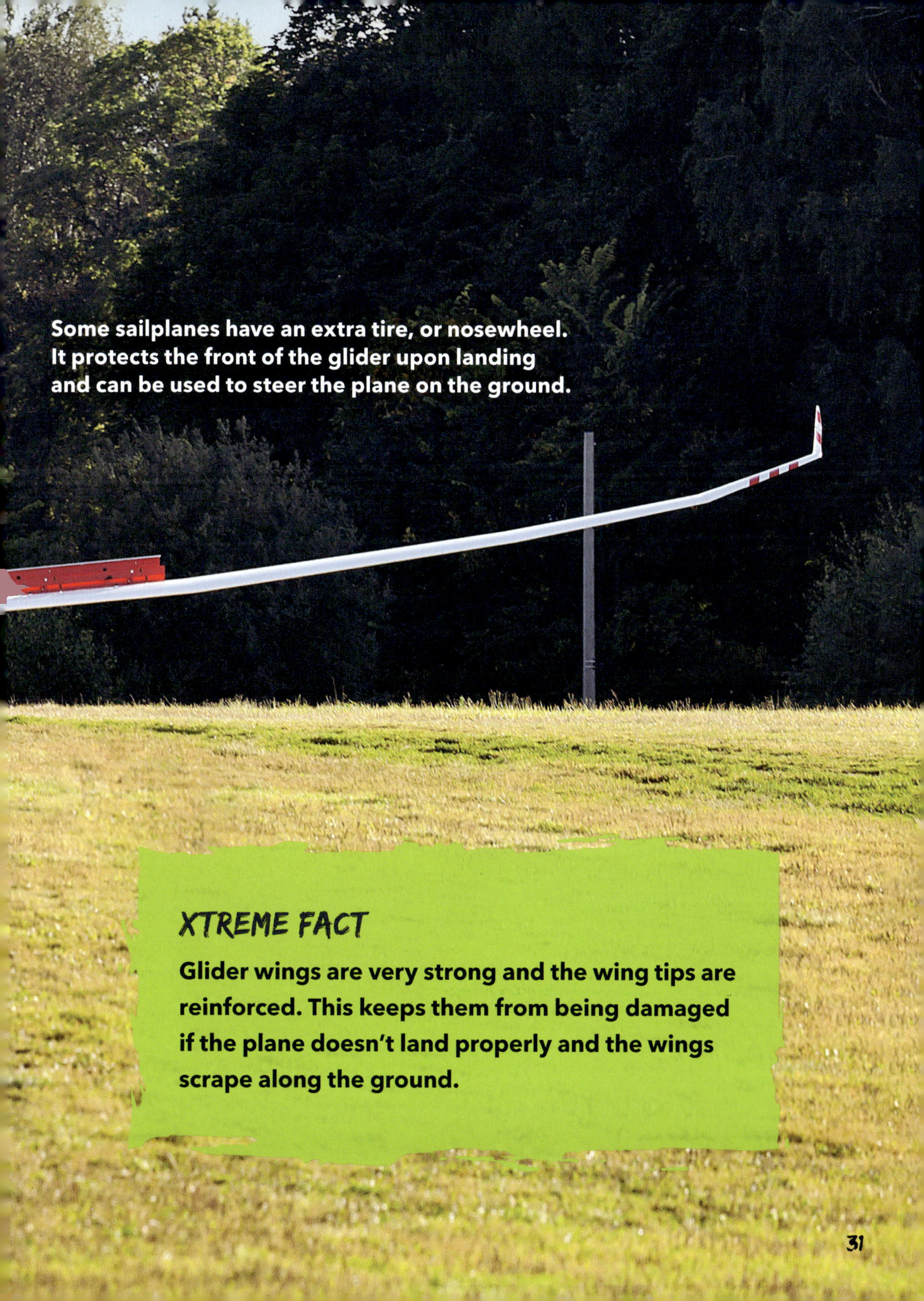

Some sailplanes have an extra tire, or nosewheel. It protects the front of the glider upon landing and can be used to steer the plane on the ground.

XTREME FACT

Glider wings are very strong and the wing tips are reinforced. This keeps them from being damaged if the plane doesn't land properly and the wings scrape along the ground.

CHAPTER 6

SAILPLANE COMPETITION CLASSES

Gliders await the start of the World Sailplane Grand Prix.

Sailplanes are divided into classes for competition. A glider's class is determined by its wingspan, weight, or whether it is a one- or two-seater.

Open Class is the oldest competition class. It includes many different types of sailplanes. Most are "big wings," but a glider may not weigh more than 850 kilograms (1,874 lbs).

A winning Open Class one-seater is the Binder EB 29R. It weighs 580 kg (1,279 lbs) and has a top speed of 280 kph (174 mph).

XTREME FACT

Up to 60 sailplanes may take part in a race. Racers often meet up near the finish. Pilots follow careful procedures with clear radio communication to keep track of each other and avoid accidents.

The ASH 30 Mi is a popular Open Class 2-seater. It has a 26.5-m (86.9-ft) wingspan and a top speed of 270 kph (168 mph).

Two-seater sailplanes allow for flying tasks to be shared by two different pilots during long flights. Comfortable cockpits are important to the pilots.

Safety and comfort are important features of the ASH 30 Mi.

Standard Class are gliders that are less expensive, smaller, and easier to fly. These sailplanes have a maximum wingspan of 15 meters (49 ft). They may not have lift-enhancing devices, such as wing **flaps**.

The Rolladen-Schneider LS-8 is a winning Standard Class plane. It has a 15-m (49-ft) wingspan and a top speed of 280 kph (174 mph).

Club Class allows older, smaller gliders to compete. Because these gliders can be very different from each other, Club Class competitions have handicapping. Scores are adjusted based on the plane's abilities.

Schweizer 1-26 is one of the most popular Club Class gliders. It was built in the United States from 1954-1979.

A Diana 2 sailplane leads other 15-m gliders in a race through South America's Andes Mountains.

Sailplanes may compete by wingspan. The 15-m Racing Class features smaller but more **maneuverable** planes. With longer wings, the 18-m Racing Class flies longer and farther than their shorter-winged counterparts.

World Class is a special glider category. In 1989, a contest was held for a sailplane that was inexpensive and safe for new pilots to fly. Poland's Warsaw Polytechnic PW-5 was named the winner in 1993.

Until 2014, anyone who took part in a World Class competition had to fly a PW-5. In 2015, the class was renamed the 13.5-m Class. Additional smaller gliders may now take part in this class, but they cannot weigh more than 300 kilograms (661 lbs).

The PW-5 is named *Smyk* (Polish for "kid"). Only about 200 of the planes were made. The top speed is 150 kph (93 mph).

CHAPTER 7

BECOMING A GLIDER PILOT

In the United States, people as young as 14 may train to **solo** in a sailplane. This requires 30-40 flights with a licensed instructor, which is about 10-12 hours of time in the air. Paragliders and hang gliders are considered **ultralight vehicles**. They do not require a license to fly.

XTREME FACT

Hang gliding is one of the world's riskiest sports. About 1 out of 1,000 hang gliders die in accidents. Safety training is important.

To get a Private Pilot-Glider certificate, a person must be at least 16 years old.

CHAPTER 8

THE FUTURE

The newest glider designs use strong, lightweight materials to make it easier for the plane to climb. Wings that can be adjusted for changing weather conditions are being developed. Electric-motor sailplanes have begun to fly. Future gliders will be safer and more **maneuverable**.

DG's Standard Class LS8e neo has an option for an electric propulsion system.

XTREME CHALLENGE

TAKE THE QUIZ BELOW AND PUT WHAT YOU'VE LEARNED TO THE TEST!

1) Gliders fly without the use of what?

2) Who invented the first fixed-wing, heavier-than-air glider?

3) What did military gliders transport during World War II?

4) What is a paraglider's wing made of?

5) How does a pilot direct a hang glider? What must he or she do to land?

6) Name five ways a sailplane may be launched.

7) What is the oldest sailplane competition class?

8) How young can a person be to solo in a sailplane? Does a person need to be licensed to fly a paraglider or hang glider? Why or why not?

GLOSSARY

fixed-wing – Wings on an aircraft that do not move. A sailplane has fixed wings.

flap – A hinged section at the back of a plane's wing that moves to different positions to help lift or land a plane.

fuselage – The main body of an aircraft. In a regular plane, the fuselage holds the pilot, crew, passengers, and cargo. A glider's fuselage is smaller as it holds only one or two people and any launching and landing equipment.

maneuverable – Able to move quickly and easily.

retract – To go back into place.

solo – To fly alone.

stall – When a glider's wings cannot produce enough lift to keep the aircraft in the air and the plane begins to drop towards the ground.

thermals – Columns of rising air created by the heating of the Earth's surface. Air near the ground expands and rises as the surface of the Earth is heated. Dark areas, such as deep green fields or blacktopped parking lots, absorb heat more quickly, which in turn heats the air above them and produces thermal air currents that gliders need to fly.

ultralight vehicle – An aircraft that is very lightweight, usually under 70 kg (154 lbs). It typically has one seat and is used for sport flying. Paragliders and hang gliders are ultralight vehicles.

World War II – A war that was fought from 1939 to 1945, involving countries around the world. The United States entered the war in December 1941.

ONLINE RESOURCES

Booklinks
NONFICTION NETWORK
FREE! ONLINE NONFICTION RESOURCES

To learn more about gliders, please visit **abdobooklinks.com** or scan this QR code. These links are routinely monitored and updated to provide the most current information available.

INDEX